A First Course
in the
Design of
Experiments
A Linear Models Approach

A First Course
in the
Design of
Experiments

A Linear Models Approach

Donald C. Weber
John H. Skillings

Miami University
Oxford, Ohio

CRC Press
Taylor & Francis Group
Boca Raton London New York

CRC Press is an imprint of the
Taylor & Francis Group, an **informa** business

CRC Press
Taylor & Francis Group
6000 Broken Sound Parkway NW, Suite 300
Boca Raton, FL 3487-2742

First issued in paperback 2020

ISBN-13: 978-0-367-57908-1 (pbk)
ISBN-13: 978-0-8493-9671-7 (hbk)

This book contains information obtained from authentic and highly regarded sources. Reasonable efforts have been made to publish reliable data and information, but the author and publisher cannot assume responsibility for the validity of all materials or the consequences of their use. The authors and publishers have attempted to trace the copyright holders of all material reproduced in this publication and apologize to copyright holders if permission to publish in this form has not been obtained. If any copyright material has not been acknowledged please write and let us know so we may rectify in any future reprint.

Visit the Taylor & Francis Web site at
http://www.taylorandfrancis.com

and the CRC Press Web site at
http://www.crcpress.com

Library of Congress Cataloging-in-Publication Data

Weber, Donald, 1925-
    A first course in the analysis of designed experiments : a linear models approach /
Donald Weber and John Skillings.
      p. cm.
    Includes bibliographical references and index.
    ISBN 0-8493-9671-9 (alk. paper)
    1. Experimental design. 2. Linear models (Statistics) I. Skillings, John H. II. Title.

QA279 .W42 1999
519.5—dc21   99-044738

Library of Congress Card Number 99-044738

PREFACE

For majors in many areas, such as mathematics, statistics, economics, engineering, and science, the first course in statistics is a course in mathematical statistics. Such a course includes much of the theory underlying the subject matter in addition to some applications. After that first course, students traditionally proceed into mainstream courses such as regression and experimental designs. These courses are typically taught with an emphasis on only methods or with an emphasis on only theory. Courses emphasizing methods largely neglect the distribution theory, which was so central to the beginning course. Theoretical courses neglect the important applications in regression and design of experiments.

This text provides a blend of the methods and theory for the important topic of design of experiments. By utilizing the linear models approach, we are able to integrate the theory with the methods and present a unified strategy for analyzing all designs. As a by-product of the linear models approach, one can obtain the traditional analysis of variance, adjusted and unadjusted sum of squares, and the f-tests. An advantage of the linear model approach is that, unlike the analysis of variance approach, the procedures are essentially the same whether the data results from designed, undesigned, unbalanced or missing cells experiments.

Features of Text

As indicated previously, the analysis of various designs is integrated in this text through the general theory of linear models. Computer usage is encouraged and illustrated throughout the text. SAS, one of the most commonly used statistical programs, is demonstrated extensively. In addition, this textbook illustrates how any regression program can also be used to carry out most analyses.

To aid in understanding the subject matter almost every section contains examples to illustrate the concepts. There are also nearly 400 exercises divided into two groups, application exercises and theoretical exercises, each with a wide range of difficulty. Those who want a more methods-oriented course may emphasize the application exercises, while the theoretical exercises may be emphasized for a more theoretical course. Data used in the exercises generally come from problems that the authors have encountered over the years in actual practice. While the data sets have often been reduced in size or otherwise modified, the reader will have the experience of working on data from real problems.

The authors have aimed to use accurate notation. For example, parameters, estimators and estimates are carefully distinguished from each other through the use of different symbols. Further, every effort is made to distinguish random variables from their realized values.

Prerequisites

Readers need to have completed a course in mathematical statistics. To fully understand and appreciate the entirety of this textbook the student should also have had exposure to matrix algebra, enough to feel comfortable with the ideas and operations. For those who have not had such an exposure to matrix algebra, there is an introduction to the topic in Appendix A of this book. Appendix A also introduces some topics that are important for this book, but are topics that are often not found in an elementary matrix algebra course. A background in regression analysis, though desirable, is not absolutely necessary. Principles of regression that are needed in this book are introduced in Chapter 6. Those who have had an introduction to regression will want to skip most of this chapter, while those without a regression background will need to cover the material. The suggested background for a reader implies that this book is intended primarily as a text for an upper-level undergraduate course or a beginning graduate course on the design of experiments.

Suggested Use

Based upon classroom testing, the authors believe that this text contains enough material to serve as at least a four semester hour course in experimental designs. There are several approaches that may be used. For a course that emphasizes both theory and applications one should first cover the linear models material in Chapters 1- 5. While most of the material in these chapters should be covered, many of the derivations should be handled outside of class to allow sufficient time to cover the applied chapters. Chapter 6 should be covered for those who have not had a course in regression. Most of the applications material in Chapters 7 - 16 may be covered, along with a few selections from Chapters 17 and 18. A course emphasizing the theory of linear models should concentrate on Chapters 1 - 5 by completing all or most of the derivations. Parts of Chapters 6 - 18 may be covered to allow for some illustrations of the linear models material. A course emphasizing applications would skim the material in Chapters 1 - 5 and then emphasize the applications material in Chapters 6 - 18. A two semester sequence emphasizing linear models in the first course (Chapters 1 - 5) and applications in the second course (Chapters 6 - 18) provides another possible use of the book.

Contents in Text

Chapter 1 provides a general overview of the topic of the design of experiments using verbal descriptions and illustrations. It presents ideas, concepts and nomenclature that the student will encounter in the remainder of the text.

In Chapters 2 through 5 the linear model is introduced with some examples. Theoretical results for the linear model, which are needed for the analysis of experimental designs, are then presented. Many examples, including numerical ones, are given throughout these chapters to illustrate the usefulness of the obtained results. Regression models and the one-way classification model are used as the primary examples in these chapters. Chapters 3 - 5 are the most theoretical in the text, while the remaining chapters are much more application orientated.

The multiple regression model and the inference formulas for this model are presented in Chapter 6. The chapter does not intend to provide a complete treatment of regression analysis, but rather it introduces the regression material that is needed later in the book.

The completely randomized design is discussed in great detail in Chapters 7 - 9. This material illustrates how the general theory for the linear model is applied to experimental design models. Special topics such as checking model assumptions, planned comparisons (Chapter 8) and multiple comparisons (Chapter 9) are introduced in this setting.

Using a format similar to that used in Chapter 7 on the completely randomized design, two other basic designs, randomized block and Latin square, are introduced and discussed in Chapters 10 - 12. The presentation flows rather easily as the theoretical foundations and analysis approach have been outlined and illustrated in earlier chapters. The ideas of adjusted versus unadjusted sum of squares, and orthogonality are introduced in this setting.

Factorial experiments, replicated in previously studied designs, are the subject matter of Chapters 13 and 14. The idea of interaction is introduced, and a brief introduction to fractional factorials closes the discussion on analyzing fixed effects factorial experiments.

In Chapter 15 the analysis of covariance is presented as a method of adjusting for the effect of extraneous factors on the response variable. Adjusted and least squares means are introduced.

Up through Chapter 15, only fixed effects experimental design models have been presented. Chapter 16 introduces the student to random and mixed effects models, the nested classification and variance components. In the final two chapters (17 and 18), these ideas and concepts are expanded upon as they are applied to special settings, such as repeated measures designs, three-stage nesting, and split-plot designs. An introduction to response surface methodology and comments on design selection conclude the work.

Acknowledgments

We wish to express our appreciation to those who have aided in the completion of this book. Our colleagues at Miami University, Drs. Robert Schaefer and Kyoungah See, and reviewers, Drs. John Stufken (Iowa State University) and Sudhir Gupta (Northern Illinois University), have provided many helpful comments that have led to an improved presentation. Ms. Jean Cavalieri has been extremely helpful with her high quality, efficient work in the preparation of this manuscript. Finally, we appreciate the patience and support provided by our spouses, Elaine and Sue, over the years.

CONTENTS

CHAPTER 1

INTRODUCTION TO THE DESIGN OF EXPERIMENTS

1.1 Designing Experiments

Experiments

Experiments are performed and observations (data) are generated in almost every discipline. Engineers perform experiments to improve the quality of products. Environmental scientists perform experiments and collect data to determine the effect of toxic substances on animals. Social scientists perform experiments to help determine the causes of certain behavior patterns.

There are many purposes for experiments. One purpose is to determine how changing the value of one variable or measurement affects another variable. For example, in making steel rods we can determine the effect of changes in the amount of a chemical A (first variable) on the tensile strength of the rod (second variable). Alternatively, we can determine the effect of increasing levels of copper in water on the lifetime of some aquatic life.

Another purpose of experiments is to compare the differences between the mean of a variable for various groups. These groups are often formed by the different values of another variable. For example, in a sex discrimination study for accountants we might want to compare the mean salary (first variable) for men and women (values of the variable sex). As another example, we could have three groups of students who were taught to read using different methods, and we might want to compare their performance on a standardized test.

Designed Experiments

A **designed experiment** is an experiment in which the experimenter plans the structure of the experiment. For example, the experimenter determines the variables to be measured, the settings for the variables, the order in which multiple trials are run, and in general sets up the experiment so that data can be obtained to help answer appropriate questions.

There are several steps that are required to set up an experiment. For most problems these steps include the ones listed below. We illustrate these various steps with the following example. A company produces metal sheets and wants to improve the strength of the sheets. The company determines that the strength of the sheets is mainly determined by two key chemicals A and B that are used in making the metal. The goal of the experiment is to study the relationship between

1

the amounts of these two chemicals and the strength of the metal sheet.

Steps for Setting Up an Experiment

- *(Goal).*   The first step is to determine the goals for the experiment. If possible, a research hypothesis should be established. For the example, the stated goal is to determine the relationship between the amount of chemical A and the amount of chemical B with the strength of the metal.

- *(Response Variable).*   Next, the **response variable** (dependent variable) for the problem should be determined. The response variable is the main variable of interest. It is the variable that is affected by the other variables in the problem. For the example, the response variable is a measurement of the strength of the metal. We are assuming that the other variables (i.e., amount of chemical A and the amount of chemical B) affect the strength.

- *(Factors).*   The **independent variables** for the problem need to be determined.   Independent variables are ones that potentially affect the response variable.   These variables will either be **controlled**, **uncontrolled** or remain **constant** in the experiment. Variables are called "controlled" if they are intentionally varied by the experimenter. These variables constitute the **factors** for the experiment. In the example, the factors are the amount of chemical A and the amount of chemical B. Since these amounts will be determined by the experimenter, these are called **controlled** variables.

Variables that the experimenter does not control are called **uncontrolled** variables. In the example there may be several uncontrolled variables.   Possibilities include temperature, humidity and other factors that might affect the strength of the metal but are not controlled by the experimenter. A diagram of this situation is given in Figure 1.1.1.

In some experiments, one or more variables are held **constant** throughout the experiment. A possibility in our example is the amount of another chemical, say C, that is to remain the same, regardless of the amount of chemical A or B.

- *(Levels).*   For each factor the experimenter must choose the different levels to be used. The **levels** of the factor are the different values that the factor assumes during the experiment. In some cases the levels are obvious and no selection needs to be made. For example, if the factor is sex, then there are clearly only two levels: female and male.

In our metal sheet example, the experimenter would need to determine how many different levels would be used for chemicals A and B. This requires knowing the "feasible" range for the chemical. In many cases two or three different levels of A will be used. For example, a high level and a low level of A might be used. Similar decisions would be required for chemical B.

If chemical A has two levels (high and low) and chemical B has three levels (high, medium and low), then the distinct test runs possible for the experiment would be (H = high, M = medium, L = low):

Test Runs

Level of chemical A: H   H   H   L   L   L

Level of chemical B: H   M   L   H   M   L

Controlled Variables

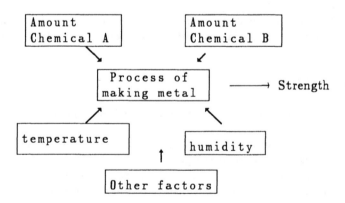

Uncontrolled Variables

Figure 1.1.1 Experiment with Controlled and Uncontrolled Variables

- *(Design)*. The experimenter now needs to choose a design for the experiment. Many decisions need to be made here including:

—Which of the test runs will be used? (All, half — which half, etc.)

—Should any test runs be duplicated? If so, how many times?

—In what order should we make the test runs? Which should be done first, last?

• *(Experiment)*. After all of this careful preparation, the experiment is run and the data is collected.

• *(Data Analysis and Conclusion)*. The data obtained from the designed experiment is now analyzed with the intent of answering the questions raised in the goals part. Conclusions are now presented with support from the data.

An article by Coleman and Montgomery (1993) provides an interesting overview of many of the issues involved in setting up an experiment.

As another example consider a study with the goal of comparing several drugs that are supposed to reduce blood pressure in humans. The response variable would be blood pressure or perhaps the amount it is reduced by the drug. Some potential independent variables would be: type drug used, subject, sex of subject, age of subject, physical condition of subject, blood pressure at the beginning of the study, other drugs a subject is taking, and many others. Important decisions that an experimenter would need to make include deciding which of these potential independent variables will be controlled as factors and which will be left uncontrolled.

Need for Good Designs

There are a number of reasons for needing a good design. Some of these are:

• We want to be sure that the data we collect is sufficient to answer the questions asked in the goal.

• The response variable varies from run to run — partly due to the different values of the controlled variables and partly due to the uncontrolled variables. We need to set up our design in such a way that the variability in response due to the uncontrolled variables (sometimes called experimental error) is not so great that it masks the effects of the controlled variables. For example, in the metal sheet problem the variation in the strength of the metal due to temperature and humidity (uncontrolled variables) could be so large that the effect on the strength due to the amounts of chemical A and B are undetectable.

• We want to design experiments that are efficient, that is, designs where we can answer the questions of interest with a minimal amount of data because of the expense associated with data collection.

1.2 Types of Designs

There are many different types of designs for experiments. Designs are classified by their characteristics. Among the ones used to classify designs are:
- The number of factors,
- The types of factors,
    - categorical versus numerical variable
    - number of levels
    - fixed versus random
- Type of randomization used, i.e., how the assignment of factor levels is made,
- Replication structure, i.e., how many times each type of test run is repeated,
- Any relationships between the factors.

To observe that more than one design is possible for a particular problem consider the following example.

Example: 1.2.1 Consider an environmental study in which the objective is to determine the level of copper in water that affects aquatic life. Daphnia, which are small aquatic animals, will be used for the study. The response variable will be the lifetime of the daphnia. The factor of interest will be the amount of copper in the water. The levels for copper will be: 0 (control group), 20 μg / liter, and 40 μg / liter. Note that a **control group** is a level of a factor that is essentially used as a base condition or a status quo condition. It is typically used for comparisons with other levels of the factor. Some uncontrolled factors include sex, physical condition of the animals, and several others. Thirty daphnia of the same age are available for the experiment. These animals will be the **experimental units**. Note that the subjects or objects on which the measurements are made are called the **experimental units**.

One possible design for this experiment is to have thirty separate chambers, one for each daphnia. We randomly assign ten daphnia to each level of copper. The resulting data structure is given in the Table 1.2.1. (Note that each * represents a single observation, i.e., a lifetime of a daphnia.)

Table 1.2.1 Experiment with Thirty Separate Chambers
Data — Lifetime

	0	*	*	*	*	*	*	*	*	*	*
Copper Level	20	*	*	*	*	*	*	*	*	*	*
	40	*	*	*	*	*	*	*	*	*	*

Another possible design for this problem could be based upon having only six separate chambers. We randomly choose two chambers for each copper level, and the thirty daphnia are randomly assigned to the six chambers subject to having five per chamber. The resulting data structure would correspond to Table 1.2.2.

Table 1.2.2 Experiment with Six Separate Chambers

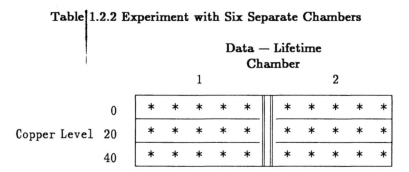

We can observe that the two designs are different. They have different structures for the data — even though the goal for the experiment remains unchanged. An experimenter will need to choose the design that best meets the needs for the experiment.

In some cases the design that is used depends upon the types of questions and analysis that are to be considered in the data analysis phase. Some typical questions or goals that could be of interest in a problem are:

- Determine whether there is a difference between the mean of the response variable for the levels of a factor.
- Determine which factor influences the response variable the most.
- Determine a functional relationship between the factors (independent variables) and the response variable.
- Determine how the factors relate to each other.
- Determine the settings for the factors that maximize or minimize the response variable.

1.3 Topics in Text

Our major focus in this text will be to discuss the issues involved in designing an experiment and analyzing the data that results from these experiments. The more commonly used designs will be discussed in great detail as will the techniques generally used in the analysis of the given data.

For each design we will present the properties of the design,